THE
LITTLE
STRAWBERRY
BOOK

Leah S. Matthews

PIATKUS

Other titles in the series

The Little Green Avocado Book
The Little Garlic Book
The Little Pepper Book
The Little Lemon Book
The Little Apple Book

©1983 Judy Piatkus (Publishers) Limited

First published in 1983 by Judy Piatkus
(Publishers) Limited of Loughton, Essex

British Library Cataloguing in Publication Data

Matthews, Leah S.
The little strawberry book.
1. Strawberries
I. Title
634'.75   SB608.S65

ISBN 0-86188-260-1

Drawings by Linda Broad
Designed by Ken Leeder

Cover photograph by John Lee

Typeset by V & M Graphics Ltd, Aylesbury, Bucks
Printed and bound by Pitman Press Ltd, Bath

CONTENTS

THE STRAWBERRY

The succulent fragrant and quite delicious-tasting strawberry is the most popular of all summer fruits. The very word strawberry conjures up nostalgic memories of long hot summer days, tea on the lawn, luxury and extravagance. When strawberries appear in the shops, summer has surely arrived.

There are many hundreds of varieties of strawberry and they are grown throughout the world. Today, the most highly sought after is the subtly-scented *fraises des bois*, wood strawberries or wild strawberries. These are beloved by gourmets everywhere, but there are many other kinds available which taste remarkably good and satisfy most palates.

Strawberries belong to the genus *Fragaria*, a member of the rose family Rosaceae. The wild strawberry, *Fragaria vesca*, is the parent of all cultivated varieties of Alpine strawberries. It grows in dry grassy places, hedgerows and woodlands. It is a low-growing perennial herb with long rooting runners from which buds arise at the nodes and tips. Its stalk and coarsely toothed small leaves are covered with soft silky hairs. White flowers, between $\frac{1}{2}$ and $\frac{3}{4}$ inch across, grow in small clusters.

The red berry or fruit is a fleshy receptacle with 'seeds' embedded closely together on the surface. After fertilisation, the receptacle enlarges and the 'seeds', which are the true fruit or achenes, become widely separated. Each achene contains a seed inside its thin dry ovary wall.

The cultivated garden strawberry, *Fragaria ananassa*, has much larger flowers and fruit than the wild strawberry. It is the result of a cross between two American species, *Fragaria chiloensis* and *Fragaria virginiana*. These plants are very different from each other; the former has glossy, dark green, leathery leaves, and the latter has non-glossy, light green, hairy, thin and coarsely-toothed leaves. The resulting hybrid has been adapted to grow under a wide range of conditions. With varying degrees of success, it can grow in irrigated dessert areas and in places receiving an annual rainfall of 100 inches; it can be found at sea level and at a height of 10,500 feet; it can survive temperatures as low as minus 50°F and as high as that of the semi tropics. However, strawberries are not truly hardy in the sense that they can withstand very low temperatures

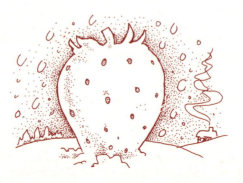

without protection. The vital part of the plant, the crown or top of the roots, lies at ground level and must be protected by snow or other cover. Without such a covering, strawberry plants are very susceptible to frost.

Most varieties are short-day plants, which means that the fruit buds are formed while the days are of medium or below-medium length. In northern latitudes, buds are formed in September and early October; further south, fruit bud formation takes place in the autumn, but growth is often resumed in early spring. This means that additional fruit buds form and the plant produces an early and a late crop.

GROWING REGIONS

The United States of America is one of the main strawberry-growing regions in the world. The United Kingdom supplies most of its own fruit during the strawberry season. Kent, Norfolk and Hampshire are the main growing areas. The fruit is also imported from many other countries, including France, Belgium, the Netherlands, Germany, Italy, Eire, Portugal, Spain, the Canary Islands, Greece, Morocco, Ethiopia, South Africa, Cyprus, Israel, New Zealand, Mexico, Chile and Pakistan.

In folklore, the strawberry plant signifies 'perfect excellence'.

WILD AND CULTIVATED VARIETIES

New varieties of strawberries are constantly being developed. The highly-prized *fraises des bois*, also known as Alpine strawberries, can sometimes be found in specialist shops. The berries are small and sweet-tasting.

A very rare wild white strawberry grows on the east coast of America and in Hawaii. It is heart-shaped, creamy white with golden seeds. Smaller than the garden or cultivated strawberry and larger than the Alpine, it is said, by those who have found it, to have a wonderful taste.

Cultivated strawberries are much larger than the wild variety. The fruit is conical in shape, with a hollow core, and when ripe the exterior is a darker red. The following list gives some of the available varieties.

ENGLISH VARIETIES

Grandee　　　　　　Early June; large fruits; good flavour.

Cambridge Rival　　Mid-June; richly flavoured but susceptible to virus.

Royal Sovereign　　Mid-June; fine flavour and ideal for dessert; susceptible to mould, mildew and virus.

Cambridge Favourite　Late June; popular with growers but not particularly tasty; better for jam-making than eating.

Red Gauntlet　　　Early July; flavour only fair, but has a second crop in autumn.

St Claude　　　　　July–October; good flavour, better in the summer than later in the year.

Gento　　　　　　August–October; needs particularly fertile soil.

AMERICAN VARIETIES

From over 70 varieties growing in America, these are among the better known.

Tioga	March–May; originated in California; good dessert quality.
Fresno	April–June; well-flavoured and suitable for dessert.
Florida Ninety	April onwards; very good dessert quality.
Headliner	May onwards; good dessert quality.
Midway	May onwards; good for eating and freezing.

THE STRAWBERRY TREE

The strawberry tree is not related to the strawberry plant, and the fruits bear no more than a faint resemblance to each other. However, the two have borne the same common name from the time of the Greeks and Romans.

The strawberry tree, *Arbutus unedo*, is a member of the Ericaceae family, to which the bilberry also belongs. Species of this evergreen tree or shrub grow in North and Central America and in the Mediterranean. In southern England it is found mainly in parks and gardens, but it grows wild and in abundance in south west Ireland.

In the wild, the tree can grow to 40 feet, though it rarely reaches more than half that height under cultivation. The bark is tinged with red and is rough, scaly and twisted. The bark of the variety *Arbutus andrachne* peels off in late spring in reddish-brown scrolls. The leaves are leathery and oval, with jagged edges and hairy stalks. At a distance, the leaves look like clusters of flowers at the tips of the twigs.

The strawberry tree flowers in September and October. The flowers are bell-shaped and creamy-white or pinkish in colour. After fertilisation the corollas drop off and in the autumn the ground beneath the tree is covered with pink or white petals.

The tree's fruit is a round berry. It is yellow, ripening to deep scarlet, with a surface that is

completely studded with little points. The berries
take about 14 months to mature and so both fruit and
flowers can be seen on the tree at the same time. The
berries are not edible until they are fully ripe, and
even then they do not suit everyone's taste. In fact,
they can be very sour indeed. The Latin word *unedo*
means 'I eat one' (only!).

DERIVATION OF 'STRAWBERRY'

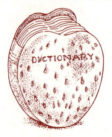

In each language, the Greeks and the Romans gave
the same common name to the strawberry plant
and the strawberry tree, presumably because of a
superficial resemblance between their fruits.

The Greek word *komaros* is taken by some to mean
the strawberry plant but it really refers to the
strawberry tree. The Greeks did not realise that the
two were not related. The Romans, who did know
the difference, nevertheless confused the two. Pliny

called the fruit the ground strawberry to distinguish it from the tree strawberry. Ovid, however, realised that they were different – he mentioned both fruits separately and made no connection between them.

Nicholas Myripsicus, a 13th-century Greek doctor, appears to have been the first person to give the strawberry a distinct name of its own. He called it 'phragouli' to distinguish it from the fruit of the *Arbutus*, the strawberry tree. By the 18th century, the present Greek word *phraouli* had entered the language.

The genus name *Fragaria* comes from the Latin, although some 16th-century botanists thought that the name came from 'fragrans' referring to the fragrance of the plant and berry. The old French word 'fresas' (modern *fraise*) and the Spanish 'fresa' derived from 'fragrans' and refer to the sweet smell.

One characteristic observed by the writers of the past which seems to have influenced the naming of the plant was its low-growing nature. This is reflected in modern names meaning earth- or ground-berry, such as the Flemish *errbesien*, Danish *jordbeer*, German *erdbeer* and Dutch *aerdbesie*.

The English word 'strawberry' comes from the Old English 'streowberie' or 'streawberige'. The plant could have received its name because of the spreading nature of its runners, which were strewn or, in Old English, 'strawed' over the ground. One reference from the 15th century is to 'strawedberry'. Alternatively, and perhaps more likely, 'straw' could refer to the manuring of the cultivated plants with straw dung.

HISTORY OF THE STRAWBERRY

In the earliest times, wild strawberries were found growing on the edges of dense forests where the soil was favourable to low-growing plants. There is little evidence of deliberate cultivation of the fruit in pre-Roman times, but Iron Age farmers had learnt to improve their crop farming methods and they may well have grown wild fruiting plants such as the strawberry.

The Romans were the first people known to have cultivated the strawberry. They were keen and innovative farmers. The species of strawberry referred to in the writings of Virgil, Ovid and Pliny was *Fragaria vesca*, the common or wood strawberry. However, little further mention of the strawberry occurs in contemporary literature for at least a thousand years. The next reference appeared in the works of the Greek doctor, Nicholas Myripsicus.

By the 14th century, at least 1,200 plants of a small-fruited indigenous European species were being cultivated in the Royal Gardens of Charles V of France, and plants grew in all French gardens at that time. Strawberries are depicted in a number of paintings and miniatures of the period. (Probably the best-known painting showing a strawberry is *The Garden of Earthly Delights* by Hieronymus Bosch, painted rather later, in 1510.)

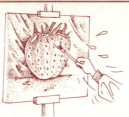

In the 15th century, strawberries were considered food for the rich. They were thought to have astringent properties and were eaten at the end of a meal in order to close up the stomach.

Ordinary people ate the fruit as and when they could obtain it. Poor children picked wild strawberries for the markets, and in London and other towns they were sold by street vendors. When there was a glut it was customary to eat well, and excessive indulgence in strawberries (as with other fruit) sometimes caused diarrhoea. This unfortunate side-effect was confused with the fluxes that often accompanied fevers, and fresh fruit was considered by many to be dangerous.

One solution to the problem of fresh fruit was to cook it. Strawberries were pulped, sieved and served in countless ways. They were used in soups or pottages; they were made into fillings for tarts; and they were turned into fools and similar creamy dishes. Sometimes the pulp was thickened with breadcrumbs or cereal flour and sweetened and spiced. Occasionally, strawberries were served with meat dishes. It is interesting to note that the redder the colour, the more appetising the strawberry was thought to be.

In England, by Elizabethan times, strawberries were being cultivated widely. Plants found growing in the woods and hedgerows were a great delicacy and they were brought into the villages and transplanted into local gardens.

> Wife into thy garden, and set me a plot,
> with strawberry roots, or the best to be got:
> Such growing abroad, among thorns in the wood,
> well chosen and picked, prove excellent good.

Shakespeare refers to strawberries in two of his plays. In *Richard III*, in order to get the Bishop of Ely to leave the room, the Duke of Gloucester asks him to send for strawberries.

> 'My lord of Ely, when I was last in Holborn
> I saw good strawberries in your garden there.
> I do beseech you send for some of them.'

In *Henry V*, Ely (again) refers to the strawberry when discussing the king.

> 'The strawberry grows underneath the nettle,
> And wholesome berries thrive and ripen best
> Neighbour'd by fruit of baser quality.'

The strawberry was not native to Europe alone. British settlers going ashore in Massachusetts in the early 17th century 'regaled themselves' with *Fragaria virginiana*, a species growing in abundance along the east coast and northern part of the continent. The Indians of New England called the fruit 'wuttahim-neash' and mixed it with meal to make bread.

According to Louis XIII of France's gardener,

Jean Rodin, the American *F. virginiana* first arrived in France (and England) in about 1624. It soon displaced most of the old European species.

Fragaria chiloensis, later to become the second ancestor (with *F. virginiana*) of today's cultivated strawberry, was a species native to south coastal Chile, the southern Cordillera of the Andes, the beaches and coastal mountains of western North America and the mountains of Hawaii. The first plants were taken from Chile to France in 1712 by Captain Frezier, an enterprising French naval officer. The original Chilean strawberries, which had been cultivated since before the arrival of the Spaniards in the 15th century, were large, white and rather flavourless.

Frezier brought only five female plants to Europe. These were given to the professor of botany at the Royal Garden in Paris, and were gradually offered by him to other European gardeners. In 1727, a plant was acquired by the Chelsea Physick Garden in London, but it remained a very rare variety. There

was a waiting period of about 100 years before *F. chiloensis* was crossed with *F. virginiana* to produce a plant bearing large, red, well-flavoured fruit.

Commercial development of strawberries for

general consumption was hindered by the perishable nature of the fruit and the slowness of transport. For a long period, cultivation was limited to areas close to large urban centres. In England, with the repeal of the Sugar Tax in 1874, jam production developed rapidly and resulted in a great increase in the quantity of soft fruit produced. Strawberries became especially popular and were grown on a much larger scale. By the beginning of the 20th century, many thousands of acres were cultivated in the Swanley area of Kent, specifically for the London markets. Fast and efficient transport systems led to the development of production centres further afield.

In America, production of strawberries expanded considerably when the Wilson variety, well suited to travel, was developed in 1851. With its introduction, strawberry-growing in New Jersey shifted from the New York area to the south of the state. After the Civil War, strawberry shipments were made to New York from the vicinity of Norfolk, Virginia. Strawberries are now grown in every state and California is the major producer.

'Strawberry leaves' is a term used to refer to a dukedom, as a ducal coronet is ornamented with eight strawberry leaves.

NUTRITIONAL VALUE OF STRAWBERRIES

Strawberries are very nutritious, containing as they do a considerable amount of Vitamin C. The Vitamin C content varies according to the variety of strawberry, the climate and harvesting conditions.

Fresh, unhulled strawberries (i.e. without the cores removed) retain their Vitamin C content for as long as they are edible, and it does not matter whether they are stored in a refrigerator or at room temperature. Hulled and punctured berries stored at room temperature will lose their Vitamin C content rapidly and become inedible within 48 hours.

Approximately 20 per cent of the Vitamin C is lost in canning, but a 4-oz portion will still supply the day's requirements. Frozen strawberries must be stored at a temperature below 0°F, otherwise all the Vitamin C is lost. Strawberries stored at between –10°F and –15°F are superior in quality and retain their nutrients better than those held at 0°F.

Strawberries are 90 per cent water. They contain various sugars, including fructose and sucrose, citric, tartaric and salicylic acids, Vitamins B_1 and B_2 and Vitamin C. Like most fruit, they are relatively rich in potassium.

Fresh strawberries have 37 calories per $3\frac{1}{2}$-oz portion, and canned strawberries, 104. The fibre content is also higher in the fresh fruit – 1.4g compared to 0.7g per $3\frac{1}{2}$-oz portion.

STRAWBERRY TIME

WHEN TO BUY STRAWBERRIES

Strawberries are in season from May until September or October, although there may be a couple of weeks in August when there are none in the shops. They are at their peak, and also at their cheapest, in July. This is the best time to buy, especially if you want to freeze a few pounds for the winter.

The popularity of strawberries means that they can be obtained all the year round from specialist shops, but if you have a craving for the fresh fruit in the winter months you will have to pay dearly for them.

BUYING STRAWBERRIES

Strawberries are so delicious and so popular that, if the quality is good and the price not extortionate, demand will be high. One of life's great pleasures is to have strawberries for the first time in a season.

Strawberries are usually displayed prominently in shops and on market stalls, and the big red luscious-looking berries sell fast. But as strawberries cannot be kept successfully for more than a day or two at room temperature, it is essential to buy from a shop with a rapid turnover.

When buying strawberries in a punnet, or small basket, examine the fruit as closely as possible to check the quality; avoid them altogether if the punnet is wrapped in plastic film as this will squash the fruit.

PICKING STRAWBERRIES

Picking your own strawberries at a strawberry farm is one of the great (but back-breaking) treats of the summer. Strawberries taste especially good if picked and eaten straight from the plant.

Strawberry plants grow low on the ground and the fruit, which ripens at different times, has to be harvested by hand. Picking the fruit is hot exhausting work, and until you have done it a few times you will have to make great effort for limited reward. Traditionally, picking was done by farmers' wives, local women, and seasonal workers, but now strawberry farms advertise for the public to come and help them pick their crops and buy the fruit at the farm shop. The method works to everyone's benefit: it saves the farmer money, and the farmer can sell direct to the public at below shop prices.

AND HAVING BOUGHT THEM …

Try to avoid hulling and washing your strawberries until you are ready to serve them. Once the leaves and hulls have been removed, the fruit can absorb water and will spoil. Store strawberries in the refrigerator and eat as soon as possible. The fresher they are, the better they will taste.

HOW TO FREEZE STRAWBERRIES

Select the strawberries for freezing when the quality is high and the price low. There are four freezing methods to choose from, but unless the berries are large and exceptionally tasty it is advisable not to freeze them whole.

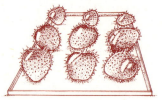

Whole fruits It is not easy, but whole fruits can sometimes be successfully frozen by the dry method. Place the strawberries slightly apart on trays and freeze until firm. Pack in rigid foil containers until required.

Whole fruits The dry sugar method gives the fruit a different flavour. Allow 4 oz caster sugar for each 1 lb strawberries. Pack the fruit in cartons, either mixed or layered with sugar. The strawberries are more likely to retain their shape if layered with sugar. When mixed, the juice is drawn out of the fruit and you can end up with strawberry purée!

Strawberry purée The easiest method is to purée the fruit before freezing. This is ideal if you want to make sorbets, ice cream, mousses, sauces, etc. You can either pass the fruit through a nylon sieve or purée it in a blender. Sweeten with 2 oz sugar to 8 oz purée and freeze in small containers.

In syrup Freeze strawberries in syrup for fruit salads. Allow approximately $\frac{1}{2}$ pint syrup for every 1 lb fruit. Dissolve 8 oz sugar in $\frac{1}{2}$ pint water, bring it to the boil, stirring, and then remove from the heat. Leave to cool. Place the fruit in a container and cover with syrup. Freeze down fast.

FREEZING TIPS

* Fruit frozen with sugar or in syrup can be stored for 9 to 12 months. Without sugar, it can be stored for only 6 to 8 months.

* To thaw strawberries, allow 6 to 8 hours in the refrigerator, or 2 to 4 hours at room temperature.

Twelve Easy Ways To Serve Strawberries

1. Put scoops of strawberry ice cream in sundae glasses, pile the strawberries on top and decorate with whipped cream and a whole strawberry.

2. Sprinkle strawberries with caster sugar and Grand Marnier and leave to stand for 2 hours. Decorate with whipped cream or serve with pouring cream.

3. Soak sliced strawberries in a mixture of fresh orange juice and curaçao. Serve with whipped cream.

4. Pour a quick strawberry sauce over vanilla ice cream and decorate with flaked almonds.

5. Place spoonfuls of vanilla and strawberry ice cream on halved bananas. Decorate with fresh strawberries and whipped cream and nuts.

6. For a special occasion, marinate strawberries in brandy and sugar – and have champagne to drink.

7. Marinate strawberries for several hours in white wine or cider.

8. Buy or make meringue boats, fill them with strawberries and top with whipped cream.

9. Fill meringue shells with strawberry purée (laced with liqueur if you are feeling extravagant) and decorate with whipped cream or ice cream. Top with a strawberry.

10. Place slices of canned pineapple in sundae dishes, scatter with fresh strawberries and serve with cream.

11. Wash the strawberries carefully but do not hull them. Dip the fruit in caster sugar and simply enjoy eating them whole.

12. Serve the strawberries with a squeeze of lemon juice to bring out the flavour.

EATING STRAWBERRIES

The strawberry ranks high among gastronomic delights. The finest dessert strawberries undoubtedly taste their best, full of natural flavour, when picked and eaten straight from the garden. Others can be made into desserts, tarts and jams.

In France, strawberries are served with burgundy or claret, or soaked in liqueurs; in England, they are offered with sugar and cream; in America, they are often cooked, turned into strawberry shortcake or served as a sundae. The Italians serve them with orange or lemon juice, and the Turks sprinkle them with lemon juice, scatter them with nuts and serve them with whipped cream flavoured with rosewater.

Curly locks, curly locks,
Wilt thou be mine?
Thou shalt not wash dishes
Nor yet feed the swine;
But sit on a cushion
And sew a fine seam,
And feed upon strawberries,
Sugar and cream.

STRAWBERRY RECIPES

STRAWBERRY SUMMER SOUP

An unusual and exceptionally delicious cold soup, perfect to serve on special occasions.

1 lb strawberries
6 fl oz white wine
3 tablespoons lemon juice
grated rind of half a lemon
2 oz sugar

Blend all the ingredients in a blender or food processor.

Transfer the soup to an attractive bowl or to individual soup dishes and chill until ready to serve.

Serves 4

'We may say of angling as Dr Butler said of strawberries, "Doubtless God could have made a better berry, but doubtless God never did".'
                                        Izaak Walton (The Compleat Angler)

STRAWBERRY SALAD

A refreshing summer lunch to eat indoors or out.

1 small lettuce
8 oz cottage cheese
8 oz strawberries, washed and halved (1 strawberry left
  whole)
4 slices pineapple, fresh or canned
4 oz green grapes, halved and seeded
4 tomatoes, thinly sliced
½ cucumber, thinly sliced
a few fresh dates (optional)

Wash the lettuce, drain well and arrange on a large platter.

Pile the cottage cheese in the centre of the lettuce and arrange the rest of the fruit attractively round it.

Place the whole strawberry on top of the cheese as a garnish.

Serves 4

Place a whole strawberry in the cavity of a canned or fresh peach or pear half and use to garnish a plate of cold meat.

Strawberry Desserts

Hot Strawberry Soufflé

This is a wonderful way to serve strawberries hot. It makes an ideal summer dinner party dessert.

butter
2 tablespoons sugar
12 oz strawberries
4 oz sugar
2 tablespoons lemon juice
½ tablespoon grated lemon rind
1 small liqueur glass brandy
4 egg yolks
4 egg whites, stiffly beaten

Grease a small casserole dish thickly with butter and sprinkle with sugar. Put in a warm place.

Purée the strawberries in a blender or crush through a nylon sieve. Place in a saucepan with the 4 oz sugar and the lemon juice and rind. Bring slowly to the boil, stirring constantly. When the mixture begins to bubble, remove from the heat and stir in the brandy. Cool for about 5 minutes.

Add the egg yolks to the strawberry mixture. Fold in the egg whites, then pour into the casserole. Cover and cook for 40 minutes at 400°F/200°C/ Gas 6.

Serves 4

STRAWBERRY SOUFFLÉ OMELETTE

This sweet omelette will not wait – it must be eaten as soon as it is ready.

8 oz strawberries
2 oz caster sugar
1 dessertspoon lemon juice
4 eggs, separated
1 oz granulated sugar
vanilla essence
1 oz butter
icing sugar

Wash, hull and slice the strawberries. Place them in a bowl with the caster sugar and lemon juice, and set on one side.

Beat the egg whites until stiff. In another bowl, beat the yolks with the granulated sugar until thick, and stir in a few drops of vanilla essence. Carefully fold the egg yolk mixture into the whites.

In a large, heavy omelette pan with a heatproof handle, melt the butter until hot. Preheat the grill.

Pour the egg mixture evenly into the pan and cook over a moderate heat until golden brown underneath and almost firm to the touch in the centre. Put the pan under the grill until the omelette sets. Spoon the strawberry mixture into the centre and fold over.

Dredge lightly with icing sugar. Serve at once.

Serves 2

Strawberry And Pineapple Compôte

8 oz sugar
¼ pint water
1 small fresh pineapple
1½ lbs strawberries, washed and hulled
juice of 1 orange
small liqueur glass Cointreau

Put the sugar and water in a pan and slowly bring to the boil, stirring to ensure that the sugar has dissolved completely.

Meanwhile, dice the pineapple and halve the strawberries. Place the fruit in a bowl with the orange juice. Pour the hot sugar syrup over the fruit, cover and leave to cool for 10 minutes before placing in the refrigerator.

Remove the compôte from the refrigerator 20 minutes before serving, and stir in the Cointreau.

Serves 6 to 8

Pile whole, hulled strawberries into hollowed grapefruit or orange shells. Serve on a platter with cheese, or offer with slices of fresh pineapple.

STRAWBERRY CREAM FOOL

Here is a simple and rather different way to use up strawberries when they are cheap but rather tasteless.

2 to 3 oz caster sugar
1 teaspoon lemon juice
1 pint strawberry purée
½ pint custard
½ pint double or whipping cream
6 small strawberries for garnish

Add the sugar and lemon juice to the strawberry purée. Fold the purée into the custard, using a metal spoon.

Whip the cream until stiff and fold into the strawberry custard, again using a metal spoon. Transfer to an attractive serving bowl or individual glasses and chill.

Decorate with small strawberries.

Serves 4 to 6

Use strawberries to decorate wedges of honeydew melon, or pile into the centre of half a cantaloupe or ogen melon.

STRAWBERRIES IN CASSIS

Cassis blends superbly with fresh strawberries and makes an exotic dessert with a tantalising flavour. You can use port wine in place of the crème de cassis and white wine, adding 2 cinammon sticks and 4 cloves.

1 lb strawberries, washed and hulled
1 lb seedless grapes or halved and seeded purple grapes
2 oz caster sugar
4 fl oz crème de cassis
white wine

In a large bowl, mix together the strawberries, grapes and sugar. Stir in the cassis and add enough wine just to cover the fruit.

Refrigerate, but remove to room temperature 10 minutes before serving.

Serves 4 to 6

For a cocktail party, cover grapefruit halves or half a white cabbage with silver foil and place, cut edge down, on a firm base. Take 6- or 8-inch wooden skewers and thread them alternately with strawberries, cheese, chunks of melon, pineapple or grapes.

FRUIT CRUSH

The fresher the fruit, the better this recipe will taste.
If you are unable to obtain ratafia biscuits, substitute
3 or 4 large macaroons, coarsely broken up.

1 packet small ratafia biscuits
1 lb strawberries, washed and hulled
8 oz raspberries
2 oz icing sugar, sieved
2 tablespoons lemon juice
¼ pint double or whipping cream, whipped

Divide the ratafia biscuits evenly between 4 sundae
dishes. Place the strawberries on top of the biscuits.

Wash the raspberries and either purée them for a
second or two in a blender or put them through a
sieve. Mix the purée with the icing sugar and lemon
juice and spoon over the strawberries, covering
as many of them as possible.

Chill until you are ready to serve, then decorate
with whipped cream.

Serves 4

Strawberry Rice Parfait

This strawberry and creamed rice sundae makes an unusual and very attractive dessert.

8 oz short-grain rice, washed
2½ tablespoons sugar
1 pint cold water
1 lb strawberries (reserving 4 for garnish)
½ pint double or whipping cream
2 tablespoons caster sugar
½ teaspoon almond essence

Place the rice, sugar and water in a pan with a tight-fitting lid and bring to the boil quickly. Stir, then lower the heat, cover and simmer for 15 minutes or until the rice is tender. Do not lift the saucepan lid during this time. When the rice is cooked, cool, then chill in the refrigerator until you are ready to prepare the meal.

Just before serving, wash and hull the strawberries. Whip the cream until stiff and fold in the caster sugar and almond essence. With a metal spoon, carefully fold the cream into the chilled rice, reserving a little cream for decoration.

Fill 4 individual sundae glasses with alternate layers of creamed rice and strawberries. Finish with whipped cream topped with a strawberry.

Serves 4

STRAWBERRY ICES
STRAWBERRY SORBET

Always a useful standby to have in the freezer, a sorbet makes a delightfully refreshing dessert to follow a heavy main course.

6 oz sugar
½ pint water
1 lb frozen strawberry purée (thawed) or fresh strawberries, puréed
2 tablespoons lemon juice
2 large egg whites, stiffly beaten

Place the sugar and water in a pan and heat, stirring, until the sugar has dissolved. Bring to the boil and simmer for 5 minutes. Remove from the heat and stir in the puréed strawberries and the lemon juice. Cool.

When cold, put the mixture in a shallow plastic container and freeze until softly frozen. Remove from the freezer and beat in a mixer until smooth, or put through a food processor. Whisk the beaten egg whites into the purée. Spoon into a container and freeze for at least 3 hours before serving.

Serves 6 to 8

STRAWBERRY ICE CREAM

This ice cream recipe is simple to make and tastes delicious. You can use either fresh strawberries or frozen and thawed strawberry purée.

12 oz strawberries, washed and hulled
6 oz caster sugar
2 tablespoons lemon juice
4 eggs, separated
½ pint double cream

In a blender or food processor, purée the strawberries with 2 oz of the sugar. Add the lemon juice.

Whisk the egg yolks in a small bowl until they are light and frothy. In a larger bowl, beat the egg whites until they are stiff, then whisk in the rest of the sugar, a spoonful at a time.

Whip the cream until it clings to the whisk when the beaters are withdrawn. (Do not over-whip as it will be difficult to fold into the strawberry mixture.)

With a metal spoon, fold the cream and the egg yolks into the egg whites. Then fold in the strawberry purée. Turn into a suitable container and freeze until firm.

Serves 8

STRAWBERRY TEAS

STRAWBERRY TARTLETS

A classic way to serve strawberries for tea. If you like decorate the tartlets with whipped cream.

Pastry
4 oz plain flour
pinch salt
2 oz butter
2 oz caster sugar
2 egg yolks

Filling
8 to 12 oz strawberries, washed and hulled
3 to 4 tablespoons redcurrant jelly

Sieve the flour and salt on to a large pastry board. Make a well in the centre and put into it the remaining ingredients. Gradually work them together until well mixed and the flour is drawn in. Knead lightly until smooth. Chill for at least an hour.

Grease 8 or 10 tartlet tins. Roll out the pastry and line the tins, pricking the bottom with a fork. Bake for 8 to 10 minutes at 350°F/180°C/Gas 4. Cool on a rack.

Arrange the strawberries in the tarts. If the strawberries are very large, cut them in half and place the cut side down. Melt the jelly in a small pan over a low heat, stirring occasionally until smooth. Spoon over the tarts and leave to set.

STRAWBERRY SHORTCAKE

No book on strawberries would be complete without a recipe for this famous dish. Here is one which is easy to make and tastes superb.

Pastry
8 oz plain flour
2 teaspoons baking powder
pinch salt
2 oz butter
2 oz sugar
4 fl oz milk

Filling
1 lb strawberries, washed and hulled
4 fl oz double or whipping cream, whipped

Sieve the flour, baking powder and salt into a bowl. Rub in the fat. Add the sugar and mix to a soft dough with the milk, adding a little more if necessary. Turn the mixture on to a board. Knead lightly and divide into two. Place each half in a greased 8-inch baking tin and flatten gradually to fit. Bake for about 20 minutes at 400°F/200°C/Gas 6. Cool in the tins, then turn out and place one round on a serving plate.

Slice all the strawberries, reserving 6 or 8 for garnish. Place a layer of strawberries on the bottom cake layer and then cover with the second round. Place the rest of the sliced strawberries on top and decorate with whipped cream and the whole berries.

CHOCOLATE-DIPPED STRAWBERRIES

An ideal treat for a rainy afternoon. Other fruit can be used.

8 oz semi-sweet chocolate
2 oz margarine
large strawberries, left whole

Melt the chocolate and margarine in the top of a double boiler set over barely simmering water.

Dip the strawberries into the chocolate one at a time and place on baking sheets lined with foil. Chill until the chocolate hardens.

Note: Make sure that the strawberries are dry before dipping, otherwise the water will cause the chocolate to curdle.

To make frosted strawberries: lightly beat an egg white until frothy. Dip whole strawberries into the egg white and then into caster sugar. Leave on a rack to dry.

STRAWBERRY FONDUE

1 lb strawberries, washed and hulled
2 tablespoons sugar
¼ teaspoon vanilla essence
2 tablespoons cognac
marshmallows

Purée the strawberries in a blender or crush through a nylon sieve. Place in a fondue pan with the sugar and vanilla essence. Stir in the cognac and heat gently.

Serve warm with marshmallows for dipping.

Serves 4

STRAWBERRY DRINKS

STRAWBERRY MILK SHAKE

4 oz strawberries, washed and hulled
½ pint cold milk
2 teaspoons caster sugar
1 scoop vanilla ice cream (optional)

Place all the ingredients in a blender. Liquidise on the fastest speed for 15 seconds.

Pour the milk shake into a tall tumbler and serve immediately, with drinking straws.

WINES TO SERVE WITH STRAWBERRIES

Sweet white wines are not to everyone's taste but there is no doubt that they do go well with desserts, and in particular with freshly-picked luscious summer strawberries. Serve the wine chilled and, if you like, pour a little over the strawberries themselves.

Highly recommended is a French sauterne or Barsac, or a German hock. Any of these will help you to capture the fresh, full flavour of the fruit.

Of course, strawberries and champagne make perfect partners. But why not try a slightly cheaper chilled sparkling wine such as Vouvray or Saumur from France, or Moselle from Germany.

STRAWBERRY DELIGHT COCKTAIL

1 measure rum
1 measure single cream
3 ice cubes
4 to 6 fresh strawberries
sugar (optional)

Mix all the ingredients in a blender and serve well chilled. If necessary, add sugar to taste.

STRAWBERRY MARGARITA COCKTAIL

1½ fl oz tequila
½ fl oz strawberry liqueur
4 oz fresh strawberries, washed and sliced
½ fl oz fresh lime juice
½ teaspoon caster sugar
1 fresh strawberry with the leaves left on

Put all the ingredients, except the single strawberry, into a blender. Add a little crushed ice and blend at low speed until the mixture is smooth. Pour into a cocktail glass and garnish with the remaining strawberry.

Serves 1

BLOODHOUND COCKTAIL

Mix ⅓ gin with ⅓ French vermouth and ⅓ Italian vermouth. Add 2 or 3 strawberries, shake well and strain.

STRAWBERRY JAM

For jam-making the strawberries must be in prime condition, sound and just-ripe.

If jam is to set properly, there must be a correct balance of pectin, acid and sugar. Some varieties of strawberries are low in pectin and acid and need one or both added. If you make a mixed jam, the pectin required by the strawberries may well be supplied by the other fruit. For plain strawberry jam, add the juice of 1 or 2 lemons to every 4 pounds of fruit, or buy bottled pectin and use as directed.

Sugar sweetens jam and also preserves it. Preserving sugar dissolves more easily than other sugars and needs little stirring. It also produces virtually no scum.

You will need a wide and deep aluminium or copper saucepan with a thick base to prevent the jam from burning. Ideally, the jam should come only half-way up the sides of the pan. You will also require jam jars, lids, a ladle and a jug. Estimate the number of jars on the basis that the finished yield will be double the weight of the sugar used.

All the equipment (apart from the saucepan, skimmer and wooden spoon) should be sterilised. If micro-organisms are left to multiply in the jam, they can spoil it. Wash everything in a weak solution of household bleach and then pour over a kettle of boiling water to wash away any taste of chlorine. Wipe with a clean tea towel or dry in a cool oven.

TESTS FOR SETTING

1. Stir the jam and use a sugar thermometer to see if the temperature has reached 221°F/105°C.
2. Spoon a little jam on to a cold saucer or plate and allow it to cool. Push your forefinger across the top of the jam and if the surface wrinkles, the jam is ready.

POTTING AND COVERING

The jars should be sterilised, dry and warm. (It is a good idea to put them in a low oven while you are cooking.) As soon as the jam has reached setting point, remove the pan from the heat. Carefully skim off any scum and leave to cool.

Pour the jam into the waiting jars and fill them to the top, using the jug, ladle or funnel. Wipe the jars and seal them immediately by placing a waxed disc, wax side down, on the jam, making sure that the disc lies flat.

When the jam is absolutely cold, put on a slightly dampened cellophane cover and secure with a rubber band. Make sure that the cellophane is not wrinkled.

Label the jars with the type of jam and the date. Store in a cool, dark, dry place.

STRAWBERRY JAM

Strawberry jam is always very popular, and a shelf-full of your own home-made preserve will also be a great source of satisfaction.

3½ lbs small strawberries, washed, dried and hulled
3 tablespoons lemon juice
3 lb sugar

Place the strawberries in a pan with the lemon juice. Mash a few strawberries to provide a little liquid and simmer gently (without adding water) until the mixture is very soft. This will take about 25 minutes.

Add the sugar and stir the mixture with a wooden spoon over a low heat until the sugar has completely dissolved. Boil rapidly until the setting point (221°F/105°C) is reached, stirring occasionally to prevent sticking. Remove any scum.

Test for setting, and then leave the jam to cool for 15 minutes. Stir to distribute the fruit and pour into warmed jars. Cover and label.

Makes approximately 5 lbs

'Like strawberry wives, that laid two or three great strawberries at the mouth of their pot, and all the rest were little ones.'

Francis Bacon, or attributed by Bacon to
Queen Elizabeth I

STRAWBERRY AND REDCURRANT JAM

For a variation use 2 pounds of gooseberries instead of redcurrants. Top and tail the gooseberries, and wash and prepare in the same way.

2 lbs redcurrants
water
4 lbs strawberries, washed and hulled
4 lbs sugar

Wash and strip the currants from their stalks. Place in a pan with a little water and simmer gently until soft. Strain through a nylon sieve to obtain the juice.

Place the strawberries in a pan with the redcurrant juice. Boil gently until the strawberries become soft. Add the sugar and stir until it has dissolved, then bring briskly to the boil. Continue to boil for about 12 minutes, or until setting point is reached. Test for setting and then leave the jam to cool for 15 minutes.

Stir. Pour into warmed jars, cover and label.

Makes approximately 7 lbs

Uncooked Strawberry Jam

If you do not wish to try your hand at preserving, here is a marvellous recipe which completely retains the fresh flavour of the fruit and yet involves no cooking. It tastes rather like a conserve.

3 lbs strawberries
3½ lbs caster sugar
8 fl oz liquid pectin
4 tablespoons lemon juice

Crush the strawberries and sugar in a bowl or purée quickly in a blender. Leave to stand at room temperature for about 1 hour, stirring occasionally.

When all the sugar has dissolved, add the pectin and lemon juice. Stir thoroughly, then pour into jars, leaving a little room at the top. Cover.

Leave at room temperature for 2 days, then refrigerate. As the mixture has not been boiled, it will spoil if kept permanently at room temperature.

Makes approximately 7 lbs

Try making your own strawberry vinegar. Add a few strawberries, fresh or frozen, to a bottle of white vinegar and let it stand for two to three weeks to mellow. Use it for a change to make dressings, marinades and sauces.

THE STRAWBERRY IN HEALTH AND MEDICINE

Strawberries were once considered to be of great medicinal value: 'a surprising remedy for the jaundice of children and particularly helping the liver of pot companions, wetters and drammers.' Another recommended use was for the treatment of fevers, and the juice of strawberries mixed with that of lemons and combined with spring water was often given to people suffering from biliousness.

However, strawberries have had periods of unpopularity. During the Middle Ages they were banished from supper and late evening meals as it was thought that eating them at night caused indigestion. In the 18th century, Horace Walpole, when discussing the serious illness of an octogenarian friend, wrote, 'Her herculean weakness, which could not resist strawberries and cream after supper, has surmounted all the ups and downs which followed her excess.'

Many fascinating Old Wives' remedies have been passed down to us over the centuries. Some may be well worth trying but check with your doctor. Some skins may prove to be allergic to strawberries so test small areas first.

* Strawberries are known to have slight laxative properties. If you have a fever, pour cold water on

crushed, bruised strawberries to make a refreshing, cooling and cleansing drink.

* Strawberries have a high assimilable iron content, so they are good for people suffering from anaemia.

* To regulate menstrual flow, acquire the habit of drinking an infusion of strawberry leaves. Taken over a long period, this drink may eventually help painful periods too. One recipe suggests infusing 3 tablespoons chopped wild strawberry leaves in $1\frac{1}{4}$ pints boiling water for 5 minutes. Strain this tea and drink a cupful, sweetened with honey if required, three times a day.

* Although strawberries have laxative qualities, the roots are useful in controlling extreme cases of diarrhoea. One cure is to boil 3 tablespoons chopped strawberry roots in $1\frac{3}{4}$ pints water for 15 minutes. Strain the concoction and drink half just before you go to sleep at night and the rest early the next morning.

* If you are prone to chilblains, try rubbing the parts likely to be affected with crushed strawberries. This treatment is said to prevent them. You will

also find it helpful to use the strawberries as a poultice, and leave them on overnight.

* Wild strawberries are said to have a beneficial effect on the body's circulation and to be especially soothing when one is feeling tense or exhausted. The leaves and roots, when dried, are particularly stimulating to the liver. A tea made by infusing the leaves or roots in boiling water is also said to act well on intestinal catarrh. One old recipe suggests mixing equal parts of the leaves of wild strawberries, blackberry leaves and woodruff with a pinch of thyme. This aromatic drink should taste good at any time of the day, and is both thirst-quenching and diuretic.

* Culpeper says that 'Strawberries cool the liver, the blood and the spleen or a hot, choleric stomach; they refresh and comfort fainting spirits and quench thirst.'

* A gargle of strawberry tea is good for you if you have ulcers or soreness in the mouth, loose teeth or spongy gums.

* An old wives' tale has it that eating fresh strawberries or drinking strawberry juice can help soothe sore or inflamed eyes.

* To cool a sunburnt face, rub it with a cut wild strawberry. Alternatively, finely sieve wild straw-berries and use the purée as a face pack. Leave on for as long as possible – at least half an hour. Wash

it off with warm water, to which have been added a few drops of simple tincture of benzoin.

* For skin inflammation and rashes, soak some chopped wild strawberry leaves in hot water for a few minutes, then put them between two pieces of muslin and use as a poultice.

STRAWBERRY NOTIONS

Strawberries conjure up such pleasant images in people's minds that the fruit is often used as a motif or logo. Here are some strawberry gift and decorating ideas well worth trying.

* Give a friend a strawberry surprise. Cover a cardboard mushroom basket with silver foil, fill it with strawberries and tie a pretty ribbon on the handle. You could always offer a large pot of thick cream, too.

* A basket of fresh strawberries makes a most attractive (and tempting) dining table decoration – just the sort of touch that will make your dinner parties memorable.

* For the person who has everything, buy a beautiful bone china strawberry dish. These dishes have a jug for cream on one side and a bowl for sugar on the other.

* Buy strawberry stencils and design your own personal greeting cards.

* Throw a strawberry party. Be bold; decorate the room (or the garden) with large cut-out cardboard strawberries; make your own invitations with a strawberry motif; buy red and white party tableware and decorate your dining table with red crêpe-paper ribbons. Make the food look as mouthwatering as possible. If the party is for children, give them strawberry milk-shakes to drink; if it is for adults, offer sparkling white wine or pink champagne! For a non-strawberry course, try and serve a dish that enhances your colour schemes. Salmon is ideal, but does depend on how wealthy you are feeling.

* Cheer up your kitchen by sticking strawberry transfers on plain walls and tiles.

* Adopt the strawberry as your personal logo. Greeting cards, postcards and stationery are all available with strawberry designs. Your friends will admire your style if any communication from you is consistent and fun.

* Make a strawberry collection. Strawberries appear in china, clay, paper, indiarubber and fabric. Look for strawberries in interesting shapes, sizes and textures.

* Sew appliqué strawberries on to your clothes – and your children's.

* Design a large strawberry-shaped cushion and fill it with kapok or foam. Scatter eye-catching satin or gingham strawberry cushions on a divan or sofa.

* A strawberry huller makes an ideal Christmas present. It is a tiny metal pincer which neatly and quickly removes the leaves and core from a strawberry without cutting into the flesh.

STRAWBERRY BEAUTY

Through the ages, women have used strawberries as a beauty aid. Isabelle of Bavaria took spring and summer baths in strawberry juice; another famous beauty, Madame Tallien, favoured a bath of crushed strawberries and raspberries, after which she was gently rubbed with sponges soaked in perfumed milk.

Today, there are several ways that strawberries can be used to make you look and feel more lovely. If you have a sensitive skin check that it is not allergic to strawberries.

* Strawberries make an excellent face pack. Mash a few ripe ones and smear the pulp on your face. Allow it to dry and, if possible, leave it on all night and wash it off in the morning with warm water. A mask of mashed, ripe strawberries will help to remove freckles and will improve the complexion.

* The juice of strawberries is said to remove tartar from the teeth. If you have no toothpaste, squash a fresh strawberry in the mouth and rub it against each tooth. It leaves a refreshing, clean taste.

* A bracing astringent lotion can be made with a batch of summer strawberries. Half-fill a container with brandy. Add strawberries, making sure that they are completely covered by the brandy. Insert a cork or rubber bung into the top of the container or screw down and leave in a warm place for a week. Strain the strawberries, preserving the brandy. (Have the strawberries that evening for supper.) Put a second batch of strawberries in the brandy, then add camphor BP (a soothing ointment for the skin, easily obtained from chemists) and leave to stand for a few more days. Strain the second batch of strawberries – but don't eat them, throw them away! Apply the lotion nightly for an improved complexion.

GROWING STRAWBERRIES

Strawberries are not difficult to grow. In order to have a constant supply of fruit throughout the season you need to have a lot of plants – up to a hundred – and a lot of space.

There are two types of strawberry plants. One-crop varieties fruit once a year in late June or early July. Perpetual or ever-bearing varieties are less hardy but produce several crops between June and October.

CHOOSING STRAWBERRY PLANTS

It is simple to establish your own runner bed, raising strawberry plants from runners. Start with young plants bought commercially and certified free from virus disease. The plants should be replaced every three years. After this time they do not crop so heavily and are more susceptible to disease.

PLANTING STRAWBERRIES

Strawberries flourish on rich, medium to heavy moisture-retaining soils, provided that they are properly drained in winter. The plants will not thrive in chalky soil.

The strawberry bed should be in a sunny open position, protected from cold winds. A south-facing plot is ideal. Dig the strawberry bed two to three weeks before planting. Double dig, remove all the weeds and add manure or garden compost to the top layer.

The amount of yield the following year depends on the date of planting. The best time to plant one-crop varieties is from July to early September. After that the crop will decrease with every week that planting is delayed. Perpetual varieties should be planted in the early spring.

Plants should be placed about 15 inches apart, in rows 30 inches apart. However, spacing depends on the variety grown and it is worth asking the nursery for advice. If the plants are sold in peat pots, soak them for an hour before planting, and plant unpotted strawberries in moist soil. Dig a hole 1 or 2 inches deeper than the roots. Put the plant on a slight mound in the centre of the hole with its roots spread out; the upper parts of the roots (the crown) should be level with the surface and only lightly covered with soil. (If the crown is buried in the soil, the crown bud may rot during the winter, and if the roots are exposed they will dry out.)

Fill in and firm the soil carefully.

For the first few weeks water the plants regularly in dry weather. Lack of water can slow growth or kill the plants.

During the autumn, clip off any runners that have grown in order to conserve the plant's energy. In the early spring, it is advisable to give the soil a dressing of sulphate of potash ($\frac{1}{2}$ oz or 1 dessertspoon to the square yard). Weed the strawberry bed whenever necessary.

At the end of April or early May, the plants will flower. On perpetual varieties, remove the flowers until late May to ensure a good late crop. Work in a mulch of peat round the crowns and under the leaves.

Towards the end of May, the first small fruits begin to appear on one-crop varieties and protection is needed to keep them from getting muddy or

mildewed. Straw is the best form of protection as it also warms the bed and helps the ripening process. Tuck straw under the trusses so that they are exposed to more sun and air. If straw is difficult to obtain, strips of polythene sheeting or strawberry mats (or collars) can be used successfully.

Remember that you will need to protect the fruit from birds. Early in the season, before they have grown, insert canes between the plants and later, when they are in full flower, fix small-mesh garden netting. Check carefully that there are no holes nor any spaces left at soil level.

Water during dry weather, and just before ripening to make the fruit swell. Once the fruit begins to ripen and the trusses are heavy enough to reach the ground, it is time to pick. Never pick wet or damp strawberries as they will rot. They must be left on the plant to dry in the sun. Picking should be carried out daily. Small, unripe fruit can be put on one side for jam-making, and the very best quality used for dessert.

Pick the strawberries with their stalks, and avoid unnecessary handling as they bruise easily.

PROPAGATION

Set aside a few healthy and high-cropping parent plants for propagation purposes. Removing the flowers will encourage the runners.

In June or July, select four runners from each of these plants. They will root easily when pegged down into the soil between the rows. Alternatively, put the largest group of leaves on each runner into small 3-inch pots filled with potting compost. Sink these full pots into the ground under the runners. Peg the runners down, and water regularly.

In about five or six weeks, when the runners are well-rooted, the small plants can be severed from the parents. Wait a further week before lifting, then transfer to a new bed with good amounts of soil attached.

STRAWBERRIES UNDER GLASS

Early crops of strawberries can be forced under glass. New plants and established runners should be trans-ferred into 6-inch pots filled with a rich growing compost mixed with manure. Leave them out until the end of November, then put them into the greenhouse as near to the glass as possible. They may be susceptible to moisture, so check the soil regularly. A moderate temperature of about 60°F/16°C is needed.

When the first fruits appear, take off any remaining flowers and increase the temperature to 65°F/18°C. As the berries begin to ripen, let the temperature rise to 80°F/27°C.

GROWING STRAWBERRIES IN BARRELS

This is an ideal method for those who have only a small sunny yard or patio. Strawberries can be grown in tubs or barrels of different sizes, and will prove decorative as well as practical.

Take an ordinary barrel and drill rows of holes 2 inches wide and 9 inches apart in the sides, through which plants can be inserted; or plant the strawberries in a half-barrel. The base of the container needs about six 1-inch holes for drainage. Containers which already have holes in them are available in the shops.

Place a layer of broken crocks or stones in the bottom of the barrel, then fill it up to the first row of holes with potting compost. Add manure. Half-fill the barrel and allow the soil to settle, then fill up with compost and manure.

Plant the strawberries through the holes from the outside. Spread out the roots and ensure that the crown of each plant is neither too shallow nor too deep. Put plants along the top of the soil too, spacing them 12 inches apart. Water when necessary. Treat the plants in the same way as those grown in beds, except that there is no need to protect them with straw.

WHAT CAN GO WRONG?

The main problems are birds, slugs, aphids and grey mould. Correct the problem with a suitable spray or chemical, except for virus disease for which there is no cure – except to re-dig the bed, buy new plants and start again.

Symptoms	*Cause*
Crippled or stunted leaves which turn yellow; small yellow, green or pink insects present.	Aphids
Leaves turn purple and curl upwards, exposing underside.	Powdery mildew
Fruits rot and become covered with a greying velvety mould.	Grey mould
Small leaves with yellow margins, in flattened tufts; plants look unhealthy.	Virus disease